Four Views of Global Warming

1) Severe Danger, 2) Mild Danger, 3) Denial, 4) Positive Event

Christophe Dupont

Light Press

ISBN: 978-1-907962-27-1

Published by Light Press

Reading, England

For Humankind

Contents

Preface

One of the things that I find most interesting about humans is their ability to disagree so much. Often people become so entrenched within their own views that they pay scant attention to the diverse range of alternative views which exist. Sometimes such disagreement is harmless, but when it comes to issues that possibly entail the very survival of the human species then such entrenched opposition has the potential to be very harmful.

Many people believe that anthropogenic global warming is a phenomenon which threatens the very survival of the human species. So, my aim is to consider a range of views concerning anthropogenic global warming in the hope that anyone who reads this book might learn a little about different perspectives on this phenomenon.

It can only be hoped that as learning increases the entrenched divisions slowly dissolve. If this were to occur then humanity could proceed in a united manner.

Introduction

The existence and nature of the phenomenon of anthropogenic global warming is something which is vociferously debated. A range of positions have been formulated concerning this phenomenon. In this book I have divided up these various positions into four distinct views.

The first view I have named the 'Severe Danger' view. According to this view anthropogenic global warming exists, it poses a serious threat to the continued existence of the human species, *and* there are no benefits arising from anthropogenic global warming (by this I mean 'benefits for the human species as a whole', rather than 'benefits for a particular group of humans' – as will be explained later). In other words, anthropogenic global warming is wholly bad and a severe threat.

The second view I have named the 'Mild Danger' view. According to this view anthropogenic global warming exists and it entails negative effects; however, these effects are 'mild' – they do not pose a serious threat to the continued existence of the human species. Furthermore, there are no benefits arising from anthropogenic global warming. In other words, anthropogenic global warming is wholly bad but the human species can adapt to the negative changes that it entails.

The third view I have named the 'Denial' view. According to this view anthropogenic global warming would be a wholly bad thing if it existed; however, it does not exist. Some advocates of this view deny the scientific basis of the phenomenon. Others accept that human actions by themselves would lead to global warming but believe that the Earth will counterbalance these human actions; this means that there is no danger from anthropogenic global warming. In short, there is nothing to be worried about.

The fourth view I have named the 'Positive Event' view. According to this view anthropogenic global warming exists, and it may be either a mild or a severe danger, but there is also a positive side to the phenomenon. In other words, whilst anthropogenic global warming presents some kind of danger to the human species there are also benefits for the human species too. Advocates of this view might believe that the benefits outweigh the danger and so believe that anthropogenic global warming is a 'good' thing.

You will notice that there is a common thread linking the thirst three positions – they all share an underlying starting assumption. These three positions all assume that human-induced global warming is solely a bad thing. This assumption arises because it is assumed that the human species exists in opposition to the rest of the planetary life-forms. That is to say, it is assumed that the human species is a parasitic species which is causing great harm to the rest of the planetary life-forms without generating any benefits for these life-forms. In short, it is believed that the selfish desire of humans to raise their living standards at the expense of non-human planetary life-forms makes the human species the enemy of these life-forms.

In contrast, the fourth view is grounded in a view according to which all of the planetary life-forms are fundamentally interconnected and not existing in opposition to each other. From this perspective the human species can be seen to be the saviour of all planetary life-forms.

In the next four chapters I consider each of these four views in turn. Then in the final chapter I draw some conclusions. In the appendix there is a paper by Neil Paul Cummins which helps to elucidate the 'Positive Event' view. I have included this appendix because the 'Positive Event' view is likely to be the view that you are least familiar with and this paper forms a useful picture of the motivations which underpin the view. I am very grateful to Neil Paul Cummins for allowing the reproduction of his paper in this book.

Chapter 1

The 'Severe Danger' View

The 'Severe Danger' view is, perhaps, the dominant view. It seems safe to say that most climate scientists (and most people?) believe that anthropogenic global warming could potentially lead to the extinction of the human species; if you believe in the *possibility* of such an eventuality then you are part of the 'Severe Danger' view.

Climate scientists work with a range of projections for the future. This range is required because nobody knows what the extent of human changes to the climate system will be in the future. In 50 years time the human species might be altering the climate system much more than it is today; alternatively, in 50 years time human alterations to the climate system might be much lower than they are today.

At one end of the scale of the projections for the future the effects of anthropogenic global warming are very serious indeed. The effects of a continued release of greenhouse gases by the human species combined with deforestation could result in a large jump in the average temperature of the Earth's atmosphere. The worry is that this jump could be the start of a chain reaction which entails a runaway greenhouse effect; the temperature of the Earth's atmosphere could shoot upwards at an accelerating rate. This would, very likely, result in the inability of the human species to live on the surface of the Earth.

It is obvious that humans – like all life-forms – need specific conditions in order to survive. So, if human actions cause the conditions of the Earth to significantly change then this could lead to the Earth becoming largely inhospitable for life. This might not mean that the human species becomes extinct because we might be able to recreate conditions which enable the continuation of the species in places such as enclosed self-sustaining capsules on the surface of the Earth, in the Earth's crust, in the atmosphere, or on places such as the Moon.

This scenario – in which humans can only survive in such self-sustaining capsules – is the extreme end of the spectrum. One can believe that anthropogenic global warming is a 'severe danger' but only have in mind far less extreme consequences. For example, one might believe that anthropogenic global warming might lead to the melting of the ice caps, which, in turn, leads to a significant rise in sea levels. The end consequences of this might be that a very large area of what is currently land is permanently submerged under water. If many of the great cities in the world today became permanently submerged under water then this might reasonably be called a 'severe outcome'. Its possibility would thus be a 'severe danger'.

Such changes might not lead to the extinction of the human species, but they would be so large as to constitute a threat to the continued existence of the human species; they could certainly change the human species as we know it today. The result of such a large area of the Earth becoming inhospitable could be the collapse of the system of states, social breakdown and a general state of anarchy.

We have been considering some of the most extreme possible outcomes that could result from anthropogenic global warming. Most people would surely accept that such extreme outcomes are a possibility. The question of importance is: How likely are these extreme outcomes to occur? The likelihood, of course, surely depends on the future actions of the human species. Although, it should be noted, many people believe that the actions that we have already undertaken make very serious changes in the future inevitable due to the effects of these actions not yet becoming fully manifest.

The short answer is that we don't know how likely it is that these extreme outcomes will obtain. What we do know is that such extreme outcomes are a possibility and that their likelihood of occurring, and their severity, is very likely going to be affected by the future activities of the human species over the next few decades and centuries.

Some people believe that such extreme outcomes are very likely to occur, if not inevitable; these people will not associate themselves with the 'Mild Danger' view. However, there are other people who whilst acknowledging the possibility of severe danger think that it is more likely that the changes that we have made, and are still making, only constitute a mild danger. In other words, one can accept that due to the limited knowledge that humans have of the climatic and oceanic systems of the Earth, and our ignorance of the extent of human activities in the future, that there could be a severe danger or that there could be a mild danger from anthropogenic global warming. These people will typically subscribe to the 'precautionary principle' and its motto that 'it is better to be safe than sorry'.

The belief in the range of possible outcomes resulting from anthropogenic global warming that I have outlined I have named the 'Severe Danger' view. This view is also characterised by the belief that anthropogenic global warming is wholly bad – there is nothing positive or good about it whatsoever.

What exactly does this mean? I should make it clear that by this I mean that at the planetary level/human species level there is nothing whatsoever positive or good about anthropogenic global warming. It might be the case that some of the people who live in a country with a relatively mild temperature might consider it to be 'good' if the temperatures there were slightly higher. However, this doesn't mean that anthropogenic global warming would be good for the Earth/life/the human species in any way whatsoever. This difference between what is good or bad for the human species as a whole, and what is good or bad for a few individual humans will periodically arise in the rest of the book.

So, according to the 'Severe Danger' view, anthropogenic global warming is wholly bad and it is an extreme threat. I stress this point because it is possible, as we will see in *Chapter Four*, to believe that whilst anthropogenic global warming is a 'severe danger' that this danger is itself a 'good' thing. That is to say, if there is both a 'benefit' and a 'danger' from anthropogenic global warming then it is possible that the 'benefit' could 'outweigh' the danger. If this were the case then whilst it would be true to say that anthropogenic global warming is a 'severe danger' it wouldn't be true to say that it is a bad thing. The 'Severe Danger' view implicitly assumes that anthropogenic global warming is wholly bad and that there is no good involved with the phenomenon whatsoever.

The advocates of the 'Severe Danger' view typically ground their beliefs in mainstream science. It is a conclusion which results from understanding a number of things. Firstly, the way that human-released greenhouse gases trap the infrared radiation from the Sun after it bounces off the Earth.

Secondly, the realisation that massive amounts of these released gases have become stored in deep ocean currents waiting to be released into the atmosphere in the future. Thirdly, the realisation that forests play a central role in regulating the atmosphere of the Earth and that the human species has deforested a very large area of the Earth. Fourthly, that once an atmospheric system goes past a 'tipping point' there can be a plethora of knock-on effects and a large jump to a very different planetary climate.

Some of these advocates will further ground their beliefs in their view of the Earth as a holistic system which is being perturbed by the human species. Such a holistic view of the Earth is often referred to as the Gaia Theory. If one sees the Earth as an interconnected whole then one has additional grounds for believing that human perturbation of both the atmosphere and the landscape of the biosphere presents a 'severe danger' to the continuation of the human species. The holistic view emphasises that the Earth is in a delicate state of balance, a state which if it is changed too much by humans can suddenly change in a radical manner to a new state of balance.

Chapter 2

The 'Mild Danger' View

The view that anthropogenic global warming is solely a severe danger is surely not the dominant view. Similarly, the view that anthropogenic global warming is solely a mild danger is surely also not the dominant view. I have suggested that the dominant view is a mixture of these two views. The dominant view being that anthropogenic global warming presents at least a mild danger with some probability (which could be anything greater than 0 per cent but less than 100 per cent) that it presents a severe danger. The dominant view is the 'Severe Danger' view which asserts that there severe danger is a possibility.

However, there are those who believe that anthropogenic global warming presents solely a mild danger (a 0 per cent probability of a severe danger). If one believes such a thing then one is part of the 'Mild Danger' view. What are the reasons why one might believe such a thing?

To be clear, the advocates of this view believe that anthropogenic global warming exists and that it entails negative effects; however, the advocates of this view believe that these effects are 'mild' – they do not pose a serious threat to the continued existence of the human species.

What kind of negative effects constitute a 'mild danger' to the human species? Some of these effects will be of a severe nature to those directly involved but not of a severe nature to the human species as a whole. So, a small increase in sea level which results in a number of low-lying islands becoming uninhabitable will not constitute a severe danger to the continuation of the human species. However, those humans who have lost their home might consider this event to be 'severe' from their personal perspective.

Other 'mild' effects will be those which can be fairly easily adapted to if the states which are affected have the required resources. If a sea level rise can be counterbalanced by an increase in the level of sea walls/improved sea defences then whilst there was a 'mild danger' there will be no negative effects (except for the financial cost involved).

Similarly, if anthropogenic climate change (which results from anthropogenic global warming) causes a small but significant increase in temperature in some countries, and a similar decrease in temperature in other countries, these effects can be relatively easily addressed if the resources are available. Adapting to these changes simply requires enough investment in air conditioning, heating equipment and other related changes.

As we have already seen, what constitutes a 'mild danger' versus a 'serious danger' will vary from one's personal perspective. So, whilst countries with lots of resources might classify a set change as a 'mild danger' less resource-rich countries might classify the same change as a 'severe danger' and this would be appropriate if their lack of resources meant that the expected effects of the change would be much more severe in their country.

I am suggesting that if a few small islands become uninhabitable that this is a 'mild danger'. However, I have claimed that if a large enough land area of the Earth became uninhabitable that this would be a 'severe danger'. So, there is clearly a grey area between what constitutes a 'mild danger' and what constitutes a 'severe danger'. That is to say, there will be a 'medium scale' land area for which it is not clear if its becoming uninhabitable constitutes a 'mild danger' or a 'severe danger'. Clearly, advocates of the 'Mild Danger' view believe that the areas of the Earth that will become uninhabitable due to anthropogenic global warming will be sufficiently small to only be worthy of the label 'mild danger'.

So, one would believe that the effects of anthropogenic global warming are only 'mild' if one believes either that the changes that the human species makes will themselves only result in relatively small changes and/or that one believes that the changes that the human species makes can relatively easily be adapted to. If the changes are relatively large but can easily be adapted to then the danger is only 'mild'. Whilst, it goes without saying, that if the changes are small then the danger will be 'mild' (we need to keep in mind that even for small changes we do not know for certain if these changes could have unforeseen consequences that are larger than one would expect given the small nature of the change which initiated them).

So, if one believes that the changes that the human species makes are relatively small given the massive scale of the non-human changes that occur on the Earth then one will have a good reason to adopt a 'Mild Danger' view. Similarly, if one believes that the human species has a great ability to adapt to the changes that arise from anthropogenic global warming then one will also have a good reason to adopt a 'Mild Danger' view.

The 'Severe Danger' view is different in that according to this view the changes that are initiated by humans are large enough to significantly alter the non-human changes that occur on the Earth. Furthermore, these human-initiated changes are on such a large scale that they cannot easily be adapted to; they are on such a scale that it makes sense to talk of a possible threat to the continuation of the human species (or, on a slightly less strong version – a possible threat to the continuation of the current way of life of the human species).

As with the 'Severe Danger' view the 'Mild Danger' view entails that there are no benefits arising from anthropogenic global warming. On this view anthropogenic global warming results in a number of changes, all of which are of a wholly negative kind, and these changes are either small or can easily be adapted to.

So, the 'Mild Danger' view is perhaps best characterised as the view that humans are not that powerful – humans have initiated changes which will lead to a small amount of global warming, but these changes pale into significance when compared to the non-human changes that occur on the Earth. In tandem, this view can be thought of as being optimistic as to the ability of humans to minimise the effects of anthropogenic global warming via adapting to these changes.

Advocates of this view might also claim that "the Earth is a robust entity". By this they mean that life on Earth has been around for billions of years successfully going about its 'business' and it is hard to imagine that a single species (humans) over the space of a relatively tiny period of time could initiate such changes as to affect the Earth in a 'severe' way. On this view there is only a mild danger from anthropogenic global warming.

Chapter 3

The 'Denial' View

You won't be surprised when I tell you that the 'Denial' view is the view that anthropogenic global warming does not exist. Advocates of this view wholeheartedly agree with the advocates of the 'Severe Danger' view and the 'Mild Danger' view that anthropogenic global warming is a danger – and a wholly bad thing – they just deny that such a phenomenon exists.

This view is obviously an extreme one. Advocates of this view obviously don't believe that anthropogenic global warming exists but that its effects will be trivial; they simply deny that the phenomenon exists. Why would one believe such a thing?

If one denies that anthropogenic global warming exists then one is clearly in a minority position. The vast majority of climate scientists, and people who are knowledgeable about the subject, believe that anthropogenic global warming does exist. However, the majority position is not always correct. There is room for alternative scientific interpretations.

Some people believe that rising carbon dioxide levels in the atmosphere do not cause an increase in the planetary temperature (as the majority believe that they do). Rather, these people claim that it is rising temperatures which cause a rising concentration of carbon dioxide in the atmosphere. If this were the case then humans would not cause global warming by releasing carbon dioxide into the atmosphere.

This view is linked to the view that changes in the temperature of the Earth are overwhelmingly caused by variations in the output of solar energy from the Sun. So, one could believe that an increase in solar energy causes an increase in the planetary temperature, and that this, in turn, causes an increase in carbon dioxide levels in the atmosphere.

Advocates of this view do not deny that the human species has, through the use of fossil fuels, taken massive amounts of previously stored carbon and released it into the atmosphere; they believe that such a release does not lead to global warming. Why would one believe such a thing? There are several possibilities.

One could simply deny the basic tenet of climate science relating to atmospheric greenhouse gases. This tenet states that higher concentrations of atmospheric greenhouse gases cause a warming of the planetary atmosphere. If one wants to reject this view one could claim that there is a relationship between the two phenomena but that the direction of causality is actually the other way round (as has already been discussed). Alternatively, one could simply deny that there is a relationship between the two phenomena.

If one denies that there is a relationship between the phenomena then one can believe that humans have increased atmospheric greenhouse gases but that this has not, and will not in the future, have any effects on the temperature of the atmosphere. Any warming can simply then be attributed to variations in incoming solar energy from the Sun.

If one believes that the direction of causality is the other way around then one can believe that increased solar energy causes an increased atmospheric temperature, and that this, in turn, causes a large increase in the level of carbon dioxide in the atmosphere. In tandem with this one can still believe that the human species has trivially increased the amount of carbon dioxide in the atmosphere. It will just be that there are two causes of the increasing atmospheric concentrations of carbon dioxide – a major one and a trivial one. This complex situation could easily lead to the 'erroneous' belief that humans by increasing atmospheric carbon dioxide levels can increase the planetary temperature.

There is another variant of the 'Denial' view. One can accept the basic tenet of climate science – that higher concentrations of atmospheric greenhouse gases cause a warming of the planetary atmosphere. One can also accept that humans have 'pumped' lots of greenhouse gases into the atmosphere. However, one could believe that the systems of the biosphere of the Earth operate in such a way that the greenhouse gases 'pumped' into the atmosphere by humans are immediately 'extracted' from the atmosphere. If this counterbalancing existed then it would clearly be the case that there would be no anthropogenic global warming resulting from the human 'pumping' of greenhouse gases into the atmosphere. On this view, rising greenhouse gas concentrations in the atmosphere would have a non-human cause (such as the increasing solar output view already discussed).

It is hard to imagine that such perfect counter-balancing could go on indefinitely, even if it existed for a while. Surely, a point would be reached when the amount of human released greenhouse gases is simply too great to be stored in non-atmospheric places? If such a point was reached then the greenhouse gases would clearly start to accumulate in the atmosphere.

Well, one could have different views on this. It all depends on how big the non-atmospheric storage areas are compared to how much greenhouse gases have been released by humans. The majority view would be that the non-atmospheric storage areas are too small, and that therefore human released greenhouse gases would ultimately accumulate in the atmosphere. However, there is almost always a minority view, and in this scenario, the minority view is that the non-atmospheric storage areas are large enough. This view would be closely aligned with the view, discussed in the previous chapter, that humans simply aren't as powerful as they think they are and that the Earth is a 'robust' entity.

So, there are various reasons why one might want to deny that human actions will lead to an increase in the average temperature of the planetary atmosphere. That is to say, there are various reasons why one might want to deny that the phenomenon of anthropogenic global warming exists. Whether any of these reasons is compelling is another matter.

Chapter 4

The 'Positive Event' View

The first three views, by definition, all entail the belief that anthropogenic global warming is wholly a 'negative event'. That is to say, at the level of the entire human species, there is nothing whatsoever positive about it. I have already outlined the important difference that exists between the level of the 'human species as whole' and 'small groups of humans'. The three views that I have already outlined entail that it is possible that mild anthropogenic global warming (if it exists) might hold some benefits for 'small groups of humans'. However, these views also entail that there would not be any benefits for the human species as a whole arising from anthropogenic global warming. That is to say, any benefits to small groups would be overwhelmingly outweighed by much more widespread negative effects.

The view which I am outlining in this chapter is very different. According to this view anthropogenic global warming does have a positive side to it at the level of the 'human species as a whole'. Furthermore, according to this view, the positive side outweighs the negative side which means that anthropogenic global warming is itself a 'Positive Event'.

The first three views I have considered are well-established and have a great number of advocates.

In contrast, this view first arose, as far as I can tell, with the publication of the 2010 book *"Is the Human Species Special?: Why human-induced global warming could be in the interests of life"* by Neil Paul Cummins. In this book Cummins outlines a particular view of the universe in which the human species came into existence with a 'purpose' to fulfil. In other words, since life arose on the Earth it has been striving to give rise to a 'human-like' species because there is something 'special' about the human species – the nature of this 'specialness' is that the human species has a special purpose to fulfil that only it can fulfil. This 'specialness' exists because the human species is the only species on the Earth that has the potential to be the saviour of life on Earth. This potential exists because the human species is that part of life which has become technological.

There are obvious merits to this technological view of the human species as the saviours of planetary life. There are a plethora of non-human threats to the continued existence of life on Earth. An enormous meteor is likely to crash into the Earth sometime in the future. This would, in all likelihood, be devastating for life on Earth. However, if life was able to become technological then it is possible that this technology could be used to deflect the meteor away from its collision course with the Earth. This is but one example of how life is more likely to survive when it becomes technological.

Many people have also realised that the continued existence of life on Earth ultimately depends on the existence of technology. This is because scientists are as certain as they can be that the Sun will eventually explode. When this day eventually comes no life will be able to exist on the Earth if the Earth remains in its present location compared to the Sun. It is possible to imagine scenarios in which life becomes so technologically advanced that it is able to move the Earth to another solar system and thereby to survive the expiration of the Sun. However, it is far more likely that life will survive this expiration by spreading out from the Earth and radiating out to the wider universe. This, of course, also requires technology of immense complexity.

There are those who seem to think that "technology is dangerous". By this they seem to have in mind a number of things. Firstly, things such as nuclear weapons and nuclear power stations, which undoubtedly have the potential to kill lots of the life-forms which exist on the Earth. Secondly, that it is technology which is the chief driving force underpinning the environmental crisis and anthropogenic global warming.

It is clearly true that technology is a 'double-edged sword'. It seems obvious to me that when technology first emerged on the Earth that this emergence was simultaneously the emergence of a set of 'dangers'. However, whilst technology clearly presents some dangers I find it hard to believe that human created technologies would ever lead to the total destruction of all life on Earth.

An astounding number of people die because of complex human technologies. Every time a person dies in a car crash this is a death resulting from the creation of complex human technologies, similarly for every airplane crash and train crash. When an astronaut dies because of a malfunction/accident this is clearly a death resulting from the creation of complex human technologies. When an electrical fire in a home or factory initiated by things such as a hairdryer, a television or a fridge, or any other electrical appliance, causes people to die these are clearly deaths resulting from the creation of complex human technologies. When people are shot, or die due to explosives, these are clearly deaths resulting from the creation of complex human technologies.

Technology clearly has an unpleasant side. However, as we have seen, without technology all life on Earth is surely doomed to ultimate extinction. If one has the interests of life on Earth as one's main concern – the continuation of the state of living which has existed on the Earth for millions of years – rather than the interests of a small segment of that life, then one will surely conclude that technology is of vital importance. Technology harms *some* life-forms but can save *life*.

How does all of this relate to anthropogenic global warming? It is claimed in *"Is the Human Species Special?: Why human-induced global warming could be in the interests of life"* that the human species is special because it is the saviour of life. This is because the human species is that part of life which has become technological. Furthermore, it is claimed that when life becomes technological it inevitably initiates an environmental crisis. In effect, the environmental crisis is an inevitable negative side-effect of the positive event which is becoming technological.

Not only does becoming technological generate an environmental crisis (which is a broad phenomenon) it also inevitable generates the specific phenomenon that is 'anthropogenic global warming'. So far, so good. One can accept that it is intelligible that life on Earth has been striving to give rise to a technological species because of the enormous overall benefits of evolving such a species. One can also see that in order to become technological a species will exploit the resources of its planet fully and thereby initiate an environmental crisis and global warming. But why is such global warming itself "in the interests of life"?

This seems to me to be the key part of, and the unique insight forwarded in *"Is the Human Species Special?: Why human-induced global warming could be in the interests of life"*. It has recently become widely accepted that life itself regulates the temperature of the Earth's atmosphere in order to maintain its temperature within the narrow range which is required for life to continue to exist (this phenomenon is often referred to as the Gaia The-ory). However, due to the increasing output of the Sun, the ability of life to regulate the temperature has been weakening for some time. The consequence of this is that in order to maintain the atmospheric temperature within the range required for life, in the future life needs to technologically regulate the atmosphere.

In short, the non-technological ability of life to regulate the temperature of the atmosphere is weakening and its maintenance requires technologi-cal regulation. This is the key component of the 'Positive Event' view – not only is technology advantageous for things such as deflecting possible meteor strikes, and not only is technology required for the ultimate survival of life, it is also required in the short-term to regulate the temperature of the Earth's atmosphere.

The last piece of the puzzle is the question of why the human species, having become technological, would use its technological expertise to regulate the temperature of the Earth's atmosphere. On the 'Positive Event' view such regulation is required if humans and other complex life-forms are to survive on the Earth, due to the phenomenon of non-anthropogenic global warming (caused by the increasing output of the Sun and the weakening homeostatic regulatory capacity of life on Earth). The answer is that humans would develop such technology if they thought that it was a necessity to deal with anthropogenic global warming. Of course, humans would also produce the technology if they fully accepted that such technology was required to deal with non-anthropogenic global warming. However, I take it that Cummins believes that non-anthropogenic global warming by itself would not provide the necessary stimulus to produce the required technology. In contrast, the phenomenon of anthropogenic global warming due to its causes and effects existing in a relatively small timescale, exacerbated by a high level of fear about the possible extent of human perturbations, could produce the necessary stimulus that is required for the technology to be produced and deployed. So, the idea is that when the effects of global warming become increasingly clear (rising sea levels, etc) then there will be a sudden realisation that the only option is to regulate the temperature of the Earth's atmosphere.

A corollary to this view is the belief that the human species is incapable of radically reducing carbon dioxide emissions. There appear to be many reasons why he believes this to be so, but the fundamental reason is clearly that it is not in our 'nature'. That is to say, the same force which led to the arising of a technological species is still in existence within the human species; it is this force which is pushing the human species towards ever greater control over its surroundings. He refers to this force as 'the force to environmental destruction'. Our inability to dramatically reduce carbon dioxide emissions actually has a purpose; it is this inability which ensures that the human species fulfils its purpose of saving planetary life.

In short, on the 'Positive Event' view, anthropogenic global warming is in the interests of life because it provides the necessary stimulus which is required for the human species to develop the atmospheric temperature controlling technology which is required if life on Earth is to survive.

On this view, as with the previous three views, there are clearly negative effects of global warming. The difference with this view is that there is also a positive side to anthropogenic global warming which outweighs the negative side. The good of life as a whole on Earth (in terms of the very survival of life on Earth) is given more importance than the negative impacts which occur to particular groups of life-forms; these negative impacts arise due to the use of technology (such as airplane crashes) and due to the effects of anthropogenic global warming.

Chapter 5

Conclusions

I have outlined four very different views of anthropogenic global warming. You will probably conclude from the outlining of these views that there is still plenty of room for debate concerning the central issues underpinning anthropogenic global warming.

There a debate as to the likely severity of the effects of anthropogenic global warming. There is a debate as to whether or not the phenomenon exists. And, there is also a debate to be had as to whether or not the phenomenon is a 'positive event' or a 'negative event'.

According to both the 'Severe Danger' view and the 'Positive Event' view anthropogenic global warming is a very severe danger. However, according to the 'Positive Event' view this danger is more than counterbalanced by the positive impacts of technology which are catalysed by anthropogenic global warming.

If one is an advocate of the 'Severe Danger' view then it is likely that one will propose measures such as recycling, reduction in resource use, renewable energy use, and possibly population size control measures, in order to deal with the severe danger (although it is possible that one might propose that the human species needs to deploy technology to directly control the temperature of the Earth's atmosphere).

However, if one is an advocate of the 'Positive Event' view then one will, in all likelihood, propose a very different solution to the severe danger faced. One will believe that the danger can only be over-come by the human species 'grabbing the bull by the horns' and taking over the non-technological ability of life to control the temperature of the Earth's atmosphere.

On the other hand, if one believes that anthro-pogenic global warming only presents a 'Mild Danger' then one will believe that the appropriate response to this danger is to adopt adaptation measures such as improved sea wall defences and the like.

Clearly, if one is an advocate of the 'Denial' view then one will not advocate any actions!

I hope that by outlining these various positions I have aided your understanding of the issues at stake in some small way. And I also hope that the following appendix has a similar beneficial effect.

Appendix

Human nature, cosmic evolution and modernity in Hölderlin

Neil Paul Cummins

Abstract:

The German Romantic Friedrich Hölderlin developed a unique perspective on the relationship between humankind and the rest of nature. He believed that humanity has a positive role to play in cosmic evolution, and that modernity is the crucial stage in fulfilling this role. In this paper I will be arguing for a reinterpretation of his ideas regarding the position of humankind in cosmic evolution, and for an application of these ideas to the 'environmental crisis' of modernity. This reinterpretation is significant because it entails an inversion of the conventional notion of causality in the 'environmental crisis'; instead of humans 'harming' nature, in the reinterpretation it is nature that causes human suffering.

Keywords:

Human Nature, Organismic Evolution, Fate, Culture, Technology, Modernity, Environmental Crisis, Human Suffering, Anthropocentricism.

Friedrich Hölderlin, one of the German Romantics, developed a distinctive viewpoint on the relationship between humankind and the rest of nature. His ideas are of particular interest because he yearned for an end to human suffering, but was also firmly convinced that humankind was inevitably destined to be separated from nature, and thereby destined to endure suffering. Hölderlin envisioned a positive role for humanity in cosmic evolution, a role which has significant implications for both human nature and cultural evolution. In this paper I will be outlining Hölderlin's ideas, and arguing for an application of them to the 'environmental crisis' of modernity. Hölderlin's conception of the human-nature relationship as part of an unfolding process of cosmological change seems to be of great relevance today, an age that is characterized by belief in the meaninglessness of human existence, and by concern about the way that we have altered the pre-human conditions of the Earth. Hölderlin's views provide a unique perspective on modernity that is worthy of serious consideration.

I start by outlining Hölderlin's views on the role of humankind in universal evolution. I then review the secondary literature on Hölderlin that relates to these ideas. I proceed to argue that Hölderlin's philosophy is applicable to, and gives a unique perspective on, the 'environmental crisis' of modernity. I argue that the existing secondary literature on Hölderlin has not recognized this, and that a reinterpretation of the role of humanity in Hölderlin's philosophy of cosmic evolution is therefore required. My central claim is that for Hölderlin, modernity and the related notion of the contemporary 'environmental crisis' is a necessary stage of cosmic evolution, and thus that it is far from a 'crisis'. Rather it is a necessary stage of disharmony that will inevitably be followed by a re-conquered harmony. I will argue that for Hölderlin this disharmony is characterized by the environmental changes that are resultant from the development of technology.

1. *Hölderlin's philosophy of human nature, cosmic evolution and modernity*

The starting point of Hölderlin's philosophy is that there must be a basic unknowable reality which precedes self-consciousness wherein subjects and objects are not in existence but are both part of a 'blessed unity of being'. He describes this unity as, "Where subject and object simply are, and not just partially, united...only there and nowhere else can there be talk of being."[1] He argues that the 'blessed unity of being' (which he also refers to as 'nature') is responsible for the coming into existence of human-

---

[1] Friedrich Hölderlin, 'Being Judgement Possibility', in J. M. Bernstein (ed.), *Classic and Romantic German Aesthetics,* Cambridge, Cambridge University Press, 2003, p. 191.

ity through using its power to initiate a division of itself into subjects and objects. This division of being causes the emergence of judgement. Hölderlin states that, "'I am I' is the most fitting example of this concept of judgement...[as] it sets itself in opposition to the *not-I,* not in opposition to *itself.*"[2]

The division means that human beings are not capable of actions that are independent of nature; Hölderlin states that, "all the streams of human activity have their source in nature."[3] It is revealing to compare this claim with the words of Hölderlin's character Hyperion, "What is man? – so I might begin; how does it happen that the world contains

---

[2] Ibid., p. 192.

[3] Alison Stone, 'Irigaray and Hölderlin on the Relation Between Nature and Culture', in *Continental Philosophy Review,* vol. 36, no. 4, 2003, p. 423.

Hölderlin

such a thing, which ferments like a chaos or moulders like a rotten tree, and never grows to ripeness? How can Nature tolerate this sour grape among her sweet clusters?"[4] For Hölderlin, man is the 'violent' being, whose coming into existence in opposition to the rest of nature was *initiated* by nature.

Hölderlin sees this opposition between man and the rest of nature as culminating in modernity – an era that he claims is characterised by the absence of the gods. In *Brot und Wein* Hölderlin writes, "Though the gods are living, Over our heads they live, up in a different world...Little they seem to care whether we live or do not."[5] A key question for

---

[4] Friedrich Hölderlin , 'Hyperion', in Eric L. Santner (ed.), *Hyperion and Selected Poems,* New York, Continuum, 1990, p. 35.
[5] Ibid., p. 185.

Hölderlin is how we deal with this separation. He envisions two possibilities – the 'Greek' response which is to dissolve the self and die, and the 'Hesperian' response of a living death.

Hölderlin came to view the 'Greek' response as hubristic, it being based on an anthropocentric desire to oppose the division initiated by nature. He thus sees the 'Hesperian' response of living and carrying out actions that are dependent on nature for their origination as the appropriate non-hubristic response to our separation. Hölderlin's position is that as nature created the separation, *only* nature can bring the separation to an end. He sees this process of separation and reconnection as part of a broader cosmic picture wherein nature is an unfolding organism rather than a huge mechanism. This organismic view enables him to envision teleological processes in nature which give rise to his claim that there will be, "eternal progress of nature towards perfection."[6]

2. Interpretations of Hölderlin and his concept of fate

In this section I set out my view of Hölderlin's conception of fate – that all human actions are part of the evolution of nature towards perfection. I do this by reviewing the existing scholarly literature on

---

[6] Ronald Peacock, *Hölderlin,* London, Methuen & Co. Ltd, 1938, p. 36.

Hölderlin and showing that whilst these interpretations all recognise parts of Hölderlin's conception of fate that they do not capture the whole of it. I start with interpretations of human nature, move on to cosmic processes, and finally consider the role of modernity within these processes.

At the level of the human there is a general consensus in the literature that Hölderlin's position is that humans are endowed by nature with qualities that shape human nature, and that this inevitably shapes human interactions with the rest of nature. There are various names in the literature for the qualities which are endowed to humans. Dennis J. Schmidt refers to the qualities present in humans as their 'formative drive.' He claims that, "Hölderlin suggests that human nature and practices are to be understood by reference to a formative drive which expresses itself as a constant need for 'art'."[7] In a similar vein, Thomas Pfau argues for an 'intellectual intuition.' He states that, "Hölderlin recasts the convergence of "freedom and necessity" as the most primordial synthesis of intellect and intuition itself, a synthesis which takes place within the subject itself. He thus approaches what Kant had repeatedly ruled out as an "intellectual intuition"."[8]

In agreement with Schmidt and Pfau, Franz Gabriel Nauen argues that for Hölderlin, "all men do

---

[7] Dennis J. Schmidt, *On Germans and Other Greeks,* Indiana University Press, 2001, p. 139.
[8] Thomas Pfau, *Friedrich Hölderlin: Essays and Letters on Theory,* New York, SUNY Press, 1988, p. 15.

in fact have the same basic character...all human activity can be derived from the same *elemental drive* in human nature."[9] The 'formative drive' / 'intellectual intuition' / 'elemental drive' identified in the literature explains why man can be seen as the 'violent' being. Human nature is to engage in 'art', to utilize the resources of nature so that culture can be generated and sustained. This generation of human culture actually benefits nature as a whole, but it requires large-scale modification of parts of non-human nature. The destiny of man is thus a disruptive one. It is clear that it is also an undesirable one. Nauen states that for Hölderlin, "Even war and

---

[9] Franz Gabriel Nauen, *Revolution, Idealism and Human Freedom: Schelling, Hölderlin and Hegel and the Crisis of Early German Idealism,* Indiana University Press, 2001, p. 139.

economic enterprise serve to fulfil the destiny of man, which is to "multiply, propel, distinguish and mix together the life of Nature".["10]

So Hölderlin sees human nature, economic production and even war as parts of a broader cosmic evolutionary process; the universe *as a whole* is seen as evolving to perfection. There will inevitably be aspects of this evolution that from a narrow perspective could be viewed as 'less than perfect'. These negative aspects of the evolutionary process – from war, to the presence of evil in its entirety – have to be seen as inescapable parts of the whole process.

---

[10] Ibid.

The key point is that for Hölderlin the cosmic evolutionary process *ends* in perfection. Thus, Ronald Peacock argues that, "the division produced by conflict is followed by a re-conquered harmony."[11] Similarly, Anselm Haverkamp argues that an interpretation of the poems *Andenken* and *Mnemosyne* is the expression, 'where danger threatens, salvation also grows.'[12] Whilst, Martin Heidegger translates the opening lines of *Patmos* as, "But where danger is, grows the saving power also."[13] Hölderlin's view is clearly that from a narrow and

---

[11] Peacock, *Hölderlin*, p. 22.

[12] Anselm Haverkamp, *Leaves of Mourning: Hölderlin's Late Work,* New York, SUNY Press, 1996, p. 48.

[13] Martin Heidegger, 'The Question Concerning Technology', in R.C. Scharff and V. Dusek (eds.), *Philosophy of Technology: The Technological Condition – An Anthology,* Oxford, Blackwell Publishing, 2003, p. 261.

short-term perspective danger and conflict are often the norm, but that these things actually play a part in bringing about a greater harmony in the future. In the long-term they are all part of the evolution of the whole universe to perfection.

Cosmic evolution is thus one long process of disharmonies and inevitably following harmonies. Peacock argues that Hölderlin's vision is of a, "harmonised process of life which comprises within itself the rhythmic movement from chaos to form and back again, and an emotional experience of this which in the sphere of nature knows only the one rapture, but in the human sphere suffering and joy."[14] It is revealing that this interpretation sees 'violent' humans as suffering, whilst nature is purely rapturous. This clearly sheds light on the question posed by Hölderlin's character Hyperion: "How can Nature tolerate this sour grape among her sweet clusters?"[15] The answer seems to be that human 'violence' *enables* nature to be rapturous. As part of this rapture humans experience suffering.

Why should suffering be a uniquely human experience? To explain this Peacock cites part of a letter from Hölderlin to his brother, "Why can they [humans] not live contented like the beasts of the field? he asks: and replies that this would be as

---

[14] Peacock, *Hölderlin*, p. 22.
[15] Hölderlin, 'Hyperion', p. 35.

unnatural in man, as in animals the tricks, or arts, man trains them to perform. Thus he establishes that the arts of man are natural to man. Culture, then, derives from nature; and the impulse to it is the characteristic which distinguishes man from the rest of creation."[16]

The human impulse to culture has culminated in the era of modernity. Hölderlin sees this period as one of great significance as he sees it as a historical epoch that is characterised by the *absence of the gods*. To be consistent with his views on harmonised evolution to perfection there must be a reason for this absence. Indeed, Peacock argues that Hölderlin thinks that, "a godless age is part of a divine mys-

---

[16] Peacock, *Hölderlin,* p. 36.

tery, it is as necessary as day, ordained by a higher power."[17] Furthermore, Heidegger claims that the gods are still present, despite their absence: "man who, even with his most exulted thought could hardly penetrate to their Being, even though, with the same grandeur as at all time, they were somehow there."[18]

The absence of the gods in modernity is deeply related to the contemporary danger that exists in modernity. It should be remembered that this danger cannot be a cause for concern for Hölderlin – as all dangers are inevitably followed by regained harmonies. Nevertheless, Heidegger attempts to

---

[17] Ibid., p. 92.
[18] Martin Heidegger, *Existence and Being,* London, Vision Press Ltd., 1956, p.190.

identify the exact danger that Hölderlin believed is present in modernity. Heidegger claims that, "the essence of technology, enframing, is the extreme danger."[19] It must follow that for Heidegger, "precisely the essence of technology must harbor in itself the growth of the saving power."[20] He sees this as occurring when the essential unfolding of technology gives rise to the possibility of opening up a "free relation" with technology which is inclusive of non-instrumental possibilities.[21]

In an interpretation of the 1802 hymn *Friedensfeier,* Richard Unger draws out Hölderlin's

---

[19] Heidegger, 'The Question Concerning Technology', p. 261.
[20] Ibid.

[21] R.C. Scharff and V. Dusek, 'Introduction to Heidegger on Technology', in R.C. Scharff and V. Dusek (eds.), *Philosophy of Technology: The Technological Condition – An Anthology,* Oxford, Blackwell Publishing, 2003, p. 248.

views on the absence of the gods in modernity.[22] In *Friedensfeier* the entire span of Western civilization is characterised as a thunderstorm which is ruled by a "law of destiny" which ensures that a certain amount of "work" is achieved. Unger argues that it is clear that this "work", "is the product of the storm itself and that it designates the harmonious totality of earthly existence during the coming era."[23] The end of the "storm" of modernity enables the arrival of a mysterious "prince" who makes it possible that, "men can now for the first time hear the "work" that

---

[22] Richard Unger, *Friedrich Hölderlin,* Boston, Twayne Publishers, 1984, pp. 100-105.
[23] Ibid., p. 102.

has been long in preparation "from morning until evening"."[24]

Following the inevitable successful accomplishment of the "work" of Western civilization, the great Spirit will disclose a Time-Image which will, "be a comprehensive depiction of the historical process and its triumphant result."[25] Unger argues that, "the Image shows that there is an alliance between the Spirit of history and the elemental divine presences of nature – for the natural elements with which man has always worked have played integral and essential parts in man's history."[26] The triumphant result of the actions of humankind in

---

[24] Ibid., p. 101.
[25] Ibid., p. 104.
[26] Ibid., p. 105.

modernity is clearly an example of a re-conquered harmony that follows division.

In Unger's interpretation of *Friedensfeier* we have a picture of modernity in which humans are carrying out "work" under a "law of destiny". The crucial factor is that humanity is ignorant that it is working under a "law of destiny" in modernity, until modernity has ended. It is then that through the Time-Image the great Spirit reveals the successful outcome of modernity, and the *nature and value* of the accomplished "work". This is a prime example of a short-term and narrow perspective entailing the perception of a lack of destiny and of needless suffering, whilst in the longer-term the same events are seen to be an inevitable part of a broader positive outcome – the evolution of the universe to perfection.

This difference of perspectives can explain an apparent contradiction in the literature between Unger's interpretation of *Friedensfeier,* and Schmidt's analysis of Hölderlin's 1801 letter to Bohlendorff. This letter was written only one year before *Friedensfeier* and Schmidt claims that in it Hölderlin's position is, "that the peculiar flow of modernity is the lack of destiny."[27] The apparently contradictory views of Unger and Schmidt can be reconciled through recalling Peacock's interpretation that, "a godless age is part of a divine mystery, it is as necessary as day, ordained by a higher

---

[27] Schmidt, *On Germans and Other Greeks,* p. 137.

power,"[28] and comparing it to Unger's claim that men are blind to the point of the "work" that they have been carrying out until the "storm" of Western civilization has passed.

The comparison reveals that the "law of destiny" applies to the activities of *humanity as a collective* in Western history, activities that are ordained by a higher power for a specific purpose. In contrast, the "lack of destiny" applies to *individual human beings*. This difference arises because individual humans are unaware that their actions are part of an inevitably unfolding cosmic plan, it is only the fruition of the plan than enables realization. Instead, humans believe that they have free will and

---

[28] Peacock, *Hölderlin,* p. 92.

live in a meaningless age. Therefore, modernity can at one and the same time be characterized as both a period governed by a "law of destiny" and a period constituted by a "lack of destiny". The difference is purely one of perspective.

This conception of modernity as simultaneously being a period of a "lack of destiny" and a "law of destiny" raises the issue of anthropocentricism. If human attitudes and actions towards nature are in the interests of nature, then it seems that there is no such thing as a *truly* anthropocentric attitude. The appropriate attitude that humans should take to the objective side of nature, given Hölderlin's philosophy, has been addressed by Alison Stone. She argues that because, "according to Hölderlin's thinking, we have become separated from nature by *its* power alone, so it is not within *our* power to undo separation."[29] Therefore, "the appropriately modest response is to endure separation – to wait, patiently, until nature may change its mode of being."[30] This means that a truly non-anthropocentric environmental view of the rest of nature requires, "the *acceptance* of disenchantment, of separation, of meaninglessness."[31]

This view is concordant with the "lack of destiny" perspective. However, when the "law of

---

[29] Stone, 'Irigaray and Hölderlin on the Relation Between Nature and Culture', p. 424.

[30] Ibid.

[31] Alison Stone, *Nature in Continental Philosophy – Week 4, Section V, Friedrich Hölderlin,* [online], http://www.lancaster.ac.uk/depts/philosophy/awaymave/408new/wk4.htm, [accessed 25 October 2005].

destiny" is taken into account, then the hidden meaning is revealed. Furthermore, the whole notion of the attitudes of individual humans then becomes irrelevant. It seems that there cannot be such a thing as a *truly* anthropocentric attitude, because all attitudes originate from nature, and they all lead to actions which fulfil the "law of destiny". It may seem that our attitudes to nature are of importance, but this is because we believe in a "lack of destiny", and are inevitably blind to the bigger picture of the "law of destiny". Whatever our attitudes as individuals, our relationship with the rest of nature as a collective would be 'for the best'.

3. A reinterpretation of the human in cosmic evolution

The interpretations of Hölderlin that I have re-
viewed all give an accurate representation of
Hölderlin's views. However, they are all partial
views. They all miss the 'big picture' of what Hölderlin's views imply about what it means to be a human
in the context of cosmic evolution, and the conse-
quent implications for the perspective from which
we should view modernity and the 'environmental
crisis'. In an attempt to fully grasp these implications I am going to defend the thesis that:
Hölderlin's philosophy leads to the conclusion that
the 'environmental crisis' is a necessary stage in the

purposeful evolution of nature towards perfection. This is an interesting thesis because, if accepted, it would supplant the conception of the meaningless-ness of human existence with a conception of positive cosmic purpose.

The argument I will be making centers on three key aspects of Hölderlin's philosophy. Firstly, that he believes that nature is purposefully evolving towards perfection. Secondly, that he believes that the achievement of this perfection requires human actions. Thirdly, that he believes that human actions are determined by nature. Acceptance of these three claims leads to the conclusion that human actions are determined by nature as a necessary stage in the purposeful evolution of nature towards perfection.

As the 'environmental crisis' of modernity is purely resultant from human actions, a second conclusion inevitably follows. This is that the 'environmental crisis' itself is determined by nature as a necessary stage in the purposeful evolution of nature towards perfection.

I will now present evidence to support the three key claims. The first claim is that Hölderlin's belief is that *nature is purposefully evolving towards perfection*. The universe can either be viewed as a giant mechanism or as an unfolding organism; Hölderlin clearly held the latter view. This conception of the universe explains his belief that nature unfolds in a way that serves its own purposes; that disharmonies are followed by regained harmonies.

This is why Peacock claims that Hölderlin believed in, "the eternal progress of nature towards perfection,"[32] and, "the emergence of perfection in the course of natural development."[33]

This firm belief clashed with Hölderlin's personal yearning for immediate perfection in life. His immense desire to see a morally just world was completely at odds with his philosophical belief that the perfection he sought could only be achieved in the course of natural development. The movement to perfection envisioned by Hölderlin is thus a fatalistic one, an inevitable evolutionary progression towards perfection. Peacock captures this with his

---

[32] Peacock, *Hölderlin,* p. 36.
[33] Ibid., p. 105.

claim that for Hölderlin there is an, "acute sense of 'Fate', of inevitability, expressed again and again in his work. Fate is revealed in the process of history... it is inherent in the passage of form to chaos, and of disintegration to a new harmony."[34]

This first claim is the most straightforward of the three. The second claim is that *Hölderlin believes that the achievement of perfection requires human actions*. The starting point in defending this claim is Hölderlin's central belief that nature *used its power* to divide itself and thereby create human-kind. This division means that the split was part of the evolutionary process rather than a random occurrence. We can ask ourselves why this may have

[34] Ibid., p. 93.

been a necessary occurrence. An initial answer seems to be Nauen's claim that, "Even war and economic enterprise serve to fulfil the destiny of man, which is to "multiply, propel, distinguish and mix together the life of Nature"."[35]

In *The Perspective from which we Have to look at Antiquity* Hölderlin asserts that, "antiquity appears altogether opposed to our primordeal drive which is bent on forming the unformed, to perfect the primordial-natural so that man, who is born for art, will naturally take to what is raw, uneducated, childlike rather than to a formed material where there has already been pre-formed [what] he wishes

---

[35] Nauen, *Revolution, Idealism and Human Freedom: Schelling, Hölderlin and Hegel and the Crisis of Early German Idealism*, p. 139.

to form."[36] In a letter to his brother he also asserts that, "the impulse to art and culture...is really a service that men render nature."[37]

The source of Hölderlin's primordeal drive to art is nature, because it is nature that created us and endowed us with our capabilities. This is clear from Peacock's interpretation that, "Man cannot be master of nature; his arts, *necessary though they may be in the scheme of things,* cannot produce the substance which they mould and transform; they can only develop the creative force, which in itself is eternal and not their work."[38]

---

[36] Friedrich Hölderlin, 'The Perspective from which We Have to Look at Antiquity', in Thomas Pfau (ed.), *Friedrich Hölderlin: Essays and Letters on Theory,* New York, SUNY Press, 1988, p. 39.
[37] Peacock, *Hölderlin,* p. 37.
[38] Ibid.

Hölderlin's primordeal drive to art in humans has inevitably led to the epoch of modernity. Human actions in this epoch appear to be central to the achievement of perfection. Hölderlin claims that modernity is an epoch that, "is as necessary as day, ordained by a higher power."[39] Furthermore, humans have been involved in "work" in modernity that is clearly constitutive of the importance of the epoch. This is clear from Unger's interpretation of *Friedensfeier* in which the "law of destiny" ensures that a certain amount of human "work" is done. The crucial factor is that humanity is ignorant that it is working under a "law of destiny" in modernity, until

---

[39] Ibid., p. 92.

modernity has ended. It is then that through the Time-Image the great Spirit reveals the successful outcome of modernity, and the nature and value of the accomplished "work".

There is no doubt that in Hölderlin's view human actions and their resultant "work" in modernity are part of purposeful evolution to perfection. What is interesting is the exact nature of the "work". There is an obvious connection between the "work" of modernity (*Friedensfeier*) and the "danger" we face in modernity (*Patmos*). Heidegger's interpretation of *Patmos* that, "the essence of technology, enframing, is the extreme danger,"[40] makes it clear that the "work" of modernity is the development of technol-

---

[40] Heidegger, 'The Question Concerning Technology', p. 261.

ogy. In fact, technological development in modernity seems to be the culmination of Hölderlin's primordeal drive to art. Furthermore, it is very hard to think of any other distinctive aspects of modernity that are resultant from human actions, present an extreme danger, and have cosmic significance. Therefore, for Hölderlin, the achievement of perfection seems to require the human development of technology.

It is interesting that Heidegger sees the danger we face from the "work" of modernity as the essence of technology rather than actual technology. Andrew Feenberg has criticised Heidegger for this abstract concentration on essences rather than the actual technology itself.[41] A "Feenberg interpretation" of *Patmos* seems to be more in accordance with Hölderlin's views than the "Heidegger interpretation", as Hölderlin's philosophy is grounded in actualities rather than essences. Hölderlin sees a positive role for actual technology in cosmic evolution; this means that *actual technology* has a cosmic purpose. Therefore, it seems that both the danger we face, and the saviour, must be the *actual* technology developed by human actions.

The importance of the human split from the rest of nature can also be seen in the words of Hölderlin's character *Hyperion:* "How should I

---

[41] Andrew Feenberg, 'Critical Evaluation of Heidegger and Borgmann', in R.C. Scharff and V. Dusek (eds.), *Philosophy of Technology: The Technological Condition – An Anthology,* Oxford, Blackwell Publishing, 2003, pp. 327-337.

escape from the union that binds all things together? We part only to be more intimately one, more divinely at peace with all, with each other. We die that we may live."[42] Human actions are thus depicted as a 'living death' that is necessary for the life (and continued movement to perfection) of nature as a whole. This explains Peacock's interpretation that, "the sphere of nature knows only the one rapture, but in the human sphere [there is] suffering and joy."[43]

The third claim is that *Hölderlin believes that human actions are determined by nature.* There are many passages in Hölderlin's novel *Hyperion* that

---

[42] Hölderlin, 'Hyperion', p. 123.
[43] Peacock, *Hölderlin*, p. 22.

attribute the responsibilities for human actions to a power or god: "There is a god in us who guides destiny as if it were a river of water, and all things are his element."[44]....."oh forgive me, when I am compelled! I do not choose; I do not reflect. There is a power in me, and I know not if it is myself that drives me to this step."[45]....."I once saw a child put out its hand to catch the moonlight; but the light went calmly on its way. So do we stand trying to hold back everchanging Fate. Oh, that it were possible but to watch it as peacefully and meditatively as we do the circling stars."[46]....."Man can change nothing and the light of life comes and departs as it

---

[44] Hölderlin, 'Hyperion', p. 11.
[45] Ibid., p. 79.
[46] Ibid., p. 22.

will."[47].....“We speak of our hearts, of our plans, as if they were ours; yet there is a power outside of us that tosses us here and there as it pleases until it lays us in the grave, and of which we know not where it comes nor where it is bound."[48]

Hölderlin's belief in the lack of human free will is perhaps clearest in his claim in a letter to his mother regarding the views of Spinoza that, “one *must* arrive at his ideas if one wants to explain everything."[49] Spinoza's ideas can be summed up as, “Nature in all its aspects is governed by necessary laws, and human being no less than the rest of

---

[47] Ibid., p. 127.

[48] Ibid., p. 29.

[49] Friedrich Hölderlin, 'No.41: To his Mother', in Thomas Pfau (ed.), *Friedrich Hölderlin: Essays and Letters on Theory*, New York, SUNY Press, 1988, p. 120.

nature is determined in all its actions and passions, contrary to those who conceive of it as 'a dominion within a dominion'."[50]

In order to make abundantly clear Spinoza's - and thus Hölderlin's – views on a lack of human free will here are two quotes from Spinoza: "I say that thing is free which exists and acts solely from the necessity of its own nature...I do not place Freedom in free decision, but in free necessity."[51] And, "a stone receives from an external cause, which impels it, a certain quantity of motion, with which it will afterwards necessarily continue to move...Next,

---

[50] Moira Gatens, *Imaginary Bodies: Ethics, Power and Corporeality,* London, Routledge, 1996, p. 111.

[51] Benedict de Spinoza, 'LVIII: To Schuller', trans. A. Wolf (ed.), *The Correspondence of Spinoza,* 2nd ed., London, Frank Cass & Co. Ltd., 1966, pp. 294-5.

conceive, if you please, that the stone while it continues in motion thinks, and knows that it is striving as much as possible to continue in motion. Surely this stone, inasmuch as it is conscious only of its own effort, and is far from indifferent, will believe that it is completely free, and that it continues in motion for no other reason than because it wants to. And such is the human freedom which all men boast that they possess, and which consists solely in this, that men are conscious of their desire, and ignorant of the causes by which they are determined."[52]

Furthermore, in an interpretation of Hölderlin's *Stutgard,* Peacock argues that, "the laws of growth

---

[52] Ibid., p. 295.

govern the culture as well as the lives of men...the

one process comprehends all things and the one

rhythm manifests itself again and again...in the

progress of history; in the spiritual life of individu-

als."[53] In this vision not only human nature, but also

the evolution of culture, is seen as an inevitable

historical progression. Peacock's interpretation of

Hölderlin is that, "man's spirit is but part of the One

Spirit,"[54] which Hölderlin insists is involved in a

"movement...through successive historical genera-

tions."[55] The spirit of man is thus governed by the

larger Spirit of nature. This is the sense in which, "all

---

[53] Peacock, *Hölderlin,* p. 25.

[54] Ibid., p. 90.

[55] Ibid., p. 114.

the streams of human activity have their source in nature."[56]

The nature of the relationship between man's spirit and the Spirit of nature is made clear in the following quote from Hölderlin's character Diotima: "a *unique destiny* bore you away to solitude of spirit as waters are borne to mountain peaks."[57] This concept of individual humans having a unique destiny was the view of Johann Herder, who was one of Hölderlin's inspirations. Herder saw nature as a great current of sympathy running through all things which manifested itself in unique inner impulses within different individuals. This means that every

---

[56] Stone, 'Irigaray and Hölderlin on the Relation Between Nature and Culture', p. 423.

[57] Hölderlin, 'Hyperion', p. 122.

human has a unique calling – an original path which they ought to tread. As Herder states, "Each human being has his own measure, as it were an accord peculiar to him of all his feelings to each other."[58] Clearly, for both Herder and Hölderlin, human actions at any one time are determined in accordance with the movements of the One Spirit of nature.

I have presented evidence for the claims that for Hölderlin: *nature is purposefully evolving towards perfection, the achievement of this perfection requires human actions, and human actions are determined by nature.* Acceptance of these three

---

[58] Charles Taylor, *Sources of the Self: The Making of the Modern Identity,* Massachusetts, Harvard University Press, 1994, p. 375.

claims leads to the conclusion that human actions are determined by nature as a necessary stage in the purposeful evolution of nature towards perfection. I now briefly argue that the 'environmental crisis' of modernity is purely resultant from human actions.

The definition of an environmental problem is: "any change of state in the physical environment which is brought about by human interference with the physical environment, and has effects which society deems unacceptable in the light of its shared norms."[59] This definition encapsulates a sliding scale of environmental problems from those that are local and temporary on the one hand, to those that are

---

[59] Peter B. Sloep and Maris C.E. van Dam-Mieras, 'Science on Environmental Problems', in P. Glasbergen and A. Blowers (eds.) *Environmental Policy in an International Context: Perspectives,* Oxford, Butterworth-Heinmann, 2003, p. 42.

global and long-lasting on the other. The 'environmental crisis' as a concept has arisen because of the emergence in the last 100 years of an increasing number of environmental problems that are towards the global and long-lasting end of the scale. The 'environmental crisis' is thus purely resultant from the *human actions* which have created environmental problems that are characterised by their global reach and long-lasting nature.

This means that the above conclusion, that human actions are determined by nature as a necessary stage in the purposeful evolution of nature towards perfection, needs amending. As the 'environmental crisis' is purely resultant from human actions, it too must be part of this purposeful evolution. Therefore, the new conclusion that inevitably follows is: *the 'environmental crisis' is determined by nature as a necessary stage in the purposeful evolution of nature towards perfection.*

4. Objections to the reinterpretation

It could be objected that there are many references to human freedom in Hölderlin's work that would seem to cast doubt on the third claim. This is particularly noticeable in his novel Hyperion. For example, Hyperion states that, "without freedom all is dead."[60] However, this objection is easily an-

---

[60] Hölderlin, 'Hyperion', p. 117.

swered because these references all appear in Hölderlin's early work, and even then they are more than counterbalanced by the opposing fatalistic views that I have outlined. In his early period Hölderlin was struggling to come to terms with the conflict between his keen moral aspirations for social change on the one hand, and his belief in perfection only arising through natural development on the other. In his later work, as is clear in his endorsement of the 'Hesperian' response to our condition, he firmly accepts the powers of natural development and the determination of human actions by nature. He realizes the futility of pursuing his idealistic moral aspirations because he accepts the illusory nature of human free will.

A further objection could be made that this re-interpretation is pointless because Darwin's theory of evolution, which emerged shortly after Hölderlin's time, gives a view of evolutionary processes that is incompatible with Hölderlin's view that there was a 'blessed unity of being' prior to the arrival of humans. We now know that the emergence of the human species – and its primordeal drive to art – was preceded by four billion years of evolution of life on Earth. It can thus be argued that there was not a 'blessed unity of being' prior to the evolution of humankind.

This is exemplified by the claim of Hans Jonas that the subject-object divide opened up four billion years ago, when, "living substance, by some original act of segregation, has taken itself out of the general integration of things in the physical context, set itself over against the world, and introduced the tension of "to be or not to be" into the neutral assuredness of existence."[61] This certainly does not appear to be a pre-human 'blessed unity of being'. However, it is interesting that Jonas also sees humans as, "a 'coming to itself' of original substance."[62]

It is clear that this Darwinian based objection does not invalidate the views of Hölderlin, or the reinterpretation of them presented in this paper. In fact, not only does evolutionary theory perfectly

---

[61] Hans Jonas, *The Phenomenon of Life: Toward a Philosophical Biology,* Illinois, Northwestern University Press, 2001, p. 4.
[62] Ibid., p. xv.

complement Hölderlin's philosophy, his philosophy *needs* it. The idea that nature could use its power to instantaneously create a being as complex as a human out of the 'blessed unity of being' is hardly defensible. In the light of our knowledge today we can simply reinterpret Hölderlin as claiming that nature used its power four billion years ago to divide the 'blessed unity of being' and create a subject/object divide. As he sees nature as an unfolding and evolving organism, the divide would give rise to human subjects after a sufficient period of time. This, ""coming to itself" of original substance", as Jonas describes it, has in actuality taken approximately four billion years.

5. Conclusion

I have argued that the existing secondary literature has not grasped the full implications of Hölderlin's thought for what it means to be a human in modernity. By drawing together Hölderlin's ideas I have sought to understand his notion of the purpose of human actions, and what this purpose means for the 'environmental crisis'.

Hölderlin's conception of nature is an organism unfolding to perfection. I have argued that he sees modernity as an important stage of this unfolding, which is characterized by the development of technology through human actions. I have further argued that this means that the 'environmental crisis' of modernity – a side-effect of the development of technology – is also an inevitable stage of this unfolding; it is in the interests of nature. As nature continues to unfold, the disharmony of modernity will be succeeded by a re-conquered harmony. I have argued that Hölderlin's 'saving power' is actual technology, as this seems most consistent with his thought. Heidegger's view, that the 'saving power' is the essencing of technology, seems inconsistent with the positive role of technology in cosmic evolution that is envisioned by Hölderlin.

The reinterpretation I have outlined clearly entails an inversion of the conventional notion of causality in the 'environmental crisis' of modernity.

Humanity is conventionally pictured as harming nature. My thesis has shown that for Hölderlin it is nature that is 'harming' humanity. We have been cast aside out of the rapture of nature into a realm of suffering and self-consciousness, with the purpose of developing technology to serve the purposes of the unfolding nature of which we are a part.

We are left with the question of what our attitudes to nature should be, given this reinterpretation of what it means to be a human in cosmic evolution. The answer is simple. As nature is the source of our individual attitudes, our attitudes to nature must be in the interests of nature. Our attitudes, whether they are techno-centric, environmentalist, quietist, or nature-exploitative are all correct for us as individuals, because in the aggregate they fulfil the purpose of nature as a whole.

BIBLIOGRAPHY

Feenberg, Andrew, 'Critical Evaluation of Heidegger and Borgmann', in R.C. Scharff and V. Dusek (eds.), *Philosophy of Technology: The Technological Condition – An Anthology*, Oxford, Blackwell Publishing, 2003.

Gatens, Moira, *Imaginary Bodies: Ethics, Power and Corporeality*, London, Routledge, 1996.

Haverkamp, Anselm, *Leaves of Mourning: Hölderlin's Late Work*, New York, SUNY Press, 1996.

Heidegger, Martin, *Existence and Being,* London, Vision Press Ltd., 1956.

Heidegger, Martin, 'The Question Concerning Technology', in R.C. Scharff and V. Dusek (eds.), *Philosophy of Technology: The Techno-logical Condition – An Anthology,* Oxford, Blackwell Publishing, 2003.

Hölderlin, Friedrich, 'Being Judgement Possibility', in J. M. Bernstein (ed.), *Classic and Romantic German Aesthetics,* Cambridge, Cambridge University Press, 2003.

Hölderlin, Friedrich, 'Hyperion', in Eric L. Santner (ed.), *Hyperion and Selected Poems,* New York, Continuum, 1990.

Hölderlin, Friedrich, 'No.41: To his Mother', in Thomas Pfau (ed.), *Friedrich Hölderlin: Essays and Letters on Theory,* New York, SUNY Press, 1988.

Hölderlin, Friedrich, 'The Perspective from which We Have to Look at Antiquity', in Thomas Pfau (ed.), *Friedrich Hölderlin: Essays and Letters on Theory,* New York, SUNY Press, 1988.

Jonas, Hans, *The Phenomenon of Life: Toward a Philosophical Biology,* Illinois, Northwestern University Press, 2001.

Nauen, Franz Gabriel, *Revolution, Idealism and Human Freedom: Schelling, Hölderlin and Hegel and the Crisis of Early German Idealism,* Indiana University Press, 2001.

Peacock, Ronald, *Hölderlin,* London, Methuen & Co. Ltd, 1938.

Pfau, Thomas, *Friedrich Hölderlin: Essays and Letters on Theory,* New York, SUNY Press, 1988.

Scharff, R. C., and Dusek, V., 'Introduction to Heidegger on Technology', in R.C. Scharff and V. Dusek (eds.), *Philosophy of Technology: The Technological Condition – An Anthology,* Oxford, Blackwell Publishing, 2003.

Schmidt, Dennis J., *On Germans and Other Greeks,* Indiana University Press, 2001.

Sloep, Peter B., and Dam-Mieras, Maris C.E. van, 'Science on Environmental Problems', in P. Glasbergen and A. Blowers (eds.) *Environmental Policy in an International Context: Perspectives,* Oxford, Butterworth-Heinmann, 2003.

Spinoza, Benedict de, 'LVIII: To Schuller', trans. A. Wolf (ed.), *The Correspondence of Spinoza*, 2nd ed., London, Frank Cass & Co. Ltd., 1966.

Stone, Alison, 'Irigaray and Hölderlin on the Relation Between Nature and Culture', in *Continental Philosophy Review*, vol. 36, no. 4, 2003. Stone, Alison, *Nature in Continental Philosophy – Week 4, Section V, Friedrich Hölderlin,* [online], http://www.lancaster.ac.uk/depts/philosophy/awaymave/408new/wk4.htm [accessed 25 October 2005].

Taylor, Charles, *Sources of the Self: The Making of the Modern Identity,* Massachusetts, Harvard University Press, 1994.

Unger, Richard, *Friedrich Hölderlin,* Boston, Twayne Publishers, 1984.